THE
Hawk's Way

ALSO BY SY MONTGOMERY

The Hummingbirds' Gift

How to Be a Good Creature

Tamed and Untamed (with
    Elizabeth Marshall Thomas)

The Soul of an Octopus

Birdology

The Good Good Pig

Search for the Golden Moon Bear:
    Science and Adventure in
    Pursuit of a New Species

Journey of the Pink Dolphins

Spell of the Tiger

Walking with the Great Apes

The Curious Naturalist

The Wild Out Your Window

Children's

The Seagull and the Sea
    Captain

Becoming a Good Creature

Condor Comeback

Inky's Amazing Escape

The Snake Scientist

The Man-Eaters of Sundarbans

Encantado: Pink Dolphin of the
    Amazon

Search for the Golden Moon Bear:
    Science and Adventure in the
    Asian Tropics

The Tarantula Scientist

Quest for the Tree Kangaroo

Saving the Ghost of the Mountain:
    An Expedition Among Snow
    Leopards in Mongolia

Kakapo Rescue: Saving the World's
    Strangest Parrot

Snowball, the Dancing Cockatoo

Temple Grandin: How the Girl Who
    Loved Cows Embraced Autism
    and Changed the World

The Tapir Scientist

Chasing Cheetahs

The Octopus Scientists

The Great White Shark Scientist

Amazon Adventure: How Tiny
    Fishes Are Saving the World's
    Largest Rainforest

The Hyena Scientist

The Magnificent Migration

THE
Hawk's Way

Encounters with
Fierce Beauty

$\backsim$

SY MONTGOMERY

Photographs by Tianne Strombeck

ATRIA BOOKS

New York • London • Toronto • Sydney • New Delhi

ATRIA
BOOKS

An Imprint of Simon & Schuster, Inc.
1230 Avenue of the Americas
New York, NY 10020

An earlier version of this material appeared as a chapter
in Sy Montgomery's book *Birdology* (2010).

First Atria Books hardcover edition May 2022

ATRIA BOOKS and colophon are trademarks of Simon & Schuster, Inc.

For information about special discounts for bulk purchases,
please contact Simon & Schuster Special Sales at 1-866-506-1949 or
business@simonandschuster.com.

The Simon & Schuster Speakers Bureau can bring authors to
your live event. For more information or to book an event, contact the
Simon & Schuster Speakers Bureau at 1-866-248-3049 or visit our website at
www.simonspeakers.com.

Interior design by Dana Sloan

Manufactured in the United States of America

1 3 5 7 9 10 8 6 4 2

Library of Congress Control Number: 2022932952

ISBN 978-1-6680-0196-7
ISBN 978-1-6680-0197-4 (ebook)

In loving memory of Nancy Cowan
March 12, 1947–January 8, 2022
master falconer, wise in the way of the hawk

THE
Hawk's Way

INTRODUCTION

Inches from my face, I hold a living dinosaur.

Like his ancestors, the creature I hold on my fist is a hunter, an eater of meat. His forebears, the theropod dinosaurs, included some of the most fearsome creatures to walk the earth: Allosaurus, Velociraptor, and Tyrannosaurus. Like them, he is a bipedal predator. Like them, he possesses large finger bones, and forward-facing eyes bestowing excellent binocular vision. Like them, when he hatched out of the egg, he was covered with down. As with many of them, his baby down then gave way to feathers.

The difference is, unlike the other dinosaurs, the one before me can fly.

His name is Mahood. He's a young Harris's hawk, a species native to the American Southwest, with bold feather markings of mahogany brown, chestnut red, and white, and long yellow legs, his feet tipped in curved, obsidian talons. In August, he was transported from the breeder where he'd hatched in upstate New York to take up residence with my friend and neighbor, Henry Walters, a poet, parent, and master falconer.

Mahood and I are meeting for the first time. He has not yet learned how to hunt. Henry is trying to teach him. Henry wants Mahood to get used to being around people, which is why he's asked me to grab my falconry glove and come over.

Mahood consents to perch on my glove. But the next moment, without any warning, he turns his head, looks into my eyes, opens his yellow, razor-sharp beak, and screams, full force, into my face.

Mahood does not like me, and is not shy about announcing this. His is not a scream of fear but of fury: the voice of an angry dinosaur. All birds, we now know from fossils and DNA, are, in fact, what became of the reptiles who once ruled the earth, creatures we all used to think were extinct. That they are not is a truth that Darwin's champion, Thomas Huxley, suspected as early as 1867; he called birds "glorified reptiles." But the connection between birds and dinosaurs is impossible to miss in a raptor.

My husband, watching from a comfortable distance, is alarmed by Mahood's scream. He's used to seeing strange dogs and cats, pigs and chickens, horses, and even an octopus, relax to my touch. But I am not surprised at all by Mahood's reaction. Hawks, as I now know well, are different.

My falconry instructor, Nancy Cowan, made this clear from the start: A hawk does not want you to touch it. It does not want to be petted. Ever. Not even a hawk you have raised from a hatchling and fed from your hand. Eventually, some hawks will, under certain circumstances, consent to your touch—but they don't like it. A single mistake handling a raptor, even one you know well, may provoke it to bite you, stab its talons into your flesh, or both.

Sometimes a hawk you've worked with for months or even years will attack. Henry's previous hawk, a big female redtail, Mary, one day flew out of a tree and, instead of landing on his glove, strafed his ear, slicing through the cartilage with her outstretched talons. The upper part of his ear flopped over like a Labrador's. (Emergency room doctors braced it so it would heal upright again.) Why? We never knew. (My husband sent me out with a hard hat the next time I flew her with Henry—but I left it behind, because many hawks dislike hats and scream at you till you take it off.)

Hawks do not play by our rules. You can never assume that a hawk, even one you raised from a chick, will forgive your mistakes—sometimes a single error ruptures the relationship forever. A hawk will not come to your rescue if you're in trouble. A hawk will not comfort you if you are sad. What a falconry hawk will do, if you do everything right, is allow you to be their hunting partner—"the junior partner," Nancy is quick to point out, for the hawk, with its exquisite vision and lightning responses, is always the superior hunter.

"It's a funny kind of relationship you have with a hawk," Henry tells me weeks later. We are walking through the forest, and Mahood is keeping pace with us, flying overhead, then perching on tree limbs, looking down and keeping track of us below, what falconers call "following on." Mahood is still immature, and Henry is well aware of the responsibility he bears for nurturing this young soul. But what is the nature of the bond you can share with a raptor?

"It's confusing," says Henry. "It's love, but all mixed up with nerves and hunger and the hunt—human love, trying to keep up with superhuman things. It's not like any other relationship you have with anyone else."

If you do everything right, a hawk will allow you to act as its servant. And for this, the falconer is profoundly grateful.

The birds of prey preserve an ancient, primal wildness, conserved in their kind since the beginning of the world. And it's exactly for this reason that, more than a decade after my first experiences with falconry, which I will share in the following pages, I still come back for more. I am still learning.

Today, the birds I first flew are gone, but I have come to know their successors, and enjoy flying them. I'm thrilled Mahood is living on our street and plan to join Henry flying him often. And I am always looking overhead for raptors, listening for the wild and savage sound of their voices.

I am drawn, and expect I always will be, to the company of hawks—to be bathed, like a baptism, in the presence of their fierce, wild glory.

Just like friendships with different people, relationships with individuals of other species have lessons to teach us. We are, of course, deeply enriched simply by learning about lifeways other than our own: I love it that some creatures I've been privileged to know can taste with their skin (octopuses), see with sound (dolphins and bats), and perceive colors we can't even imagine (birds and some reptiles). But animals also have much to show us about how we, as humans, can more meaningfully and compassionately encounter the wider world.

Our fellow animals teach us lessons about the delights of

sameness and difference. They immerse us in wonder. They lead us to humility; they inspire us to reverence. They teach us the many facets of love.

The ancient Greeks said there were four kinds of love. *Eros* takes its name from the dangerous and powerful god of desire, whose Roman counterpart was the mischievous Cupid. Eros described what we in the West now call romantic love. *Storge* is a very different kind of love: the instinctual affection of a parent toward its offspring and vice versa. *Philia* is the love shared between friends and equals, and gives us words such as philanthropy.

The highest form of love was called a*gape*. This is a love untainted by expectations, a love without external reward. The other kinds of love, though celebrated and essential, are transactional. Eros yearns for sexual fulfillment. Philia's bonds are forged by mutual friendship and trust. Storge is run by selfish genes. But unlike these, agape is selfless, pure, and unconditional. In the Bible, agape came to stand for the love God has for His creation, and the love humans should endeavor to feel toward the Creator in worship. But in the way of humans, we too often make even worship transactional: giving our allegiance in exchange for eternal life, or the promise of consorting, after a glorious death, with seventy-two virgins. ("Oh Lord," sang Janis

Joplin, "won't you buy me / a Mercedes Benz?") This is not true to the spirit of agape: Unlike all the other kinds of love, agape asks nothing in return.

This is what a hawk can teach you: how to love like a god.

∽∾

Once again, on a recent September afternoon, I am holding a dinosaur. It's a blue-sky day, a light wind to the northwest, and the temperature has risen from the morning's fifty degrees to sixty-five. Perfect. With seventy-five other people, I am at Miller State Park in Peterborough, New Hampshire, standing on Pack Monadnock mountain during one of the peak weeks of the hawk migration.

My husband and I often come up here on fall days to join other hawk watchers hoping to catch a glimpse of the spectacle overhead. From September to December, nearly twenty thousand raptors of fifteen species will pass over the mountain—often in groups ranging from handfuls to hundreds called kettles, swirling and spiraling upward on rising columns of warm air—on their way to southern wintering grounds. Volunteers and Hawk Watch staff from our local Audubon Society and the Harris Center for Conservation Education tally the birds' numbers.

But today is extra special. As part of a once-a-year tradition, a crowd gathers to watch the release of hawks that had been found sick or injured the previous summer—birds that, thanks to the expert care of Henniker, New Hampshire, wildlife rehabilitator Maria Colby, are now healthy and whole, and ready to join the migration.

My friend Julie Brown, Hawk Migration Association of North America's monitoring site coordinator, is here, as usual. But to my surprise, she has pulled a young broad-winged hawk from Maria's transport carrier and has handed it to me.

The broad-wing is a smallish, stocky, forest-dwelling raptor with chocolate streaks running across a light breast, and black and white bands on the tail. I often hear this species calling when one passes overhead, a distinctive, drawn-out, one-note whistle, but I have never before held one, or seen one this close.

Julie tells me to hold the bird in a manner completely unlike anything I've done in falconry. The bird doesn't perch on my glove; instead, I hold firmly onto the yellow legs and feet with a leather-gloved hand, while using my other, ungloved hand to keep the wings from beating. Julie instructs me to hang on to the bird like this in order to display it to the admiring crowd before I let it go.

The hawk is understandably furious at this indignity. The

broad-wing swivels its neck to stare into my face. To the bird, I am a monstrosity, its captor, its evil enemy. With fearless fury, the hawk lets loose a scream that feels as if, with the help of a tiny spark, it could ignite a fireball and consume me, the crowd, the entire mountain. I am a 120-pound monster holding this one-pound creature in my hand. I am a heavy beast made of flesh and fluid; the bird I hold is made mostly of air, whose feathers outweigh its skeleton. Yet I stand deeply humbled by the power of its courage as it screams at me in fiery defiance.

So consumed am I with the broad-wing's incandescent presence that I do not realize that the person who rescued the bird, Nancy Delaney, is in the crowd today. Weeks later, I have the pleasure of meeting her. Nan is a horsewoman, about my age, with short auburn hair and brown eyes. She has always loved animals. But she has never had anything to do with hawks.

On a rainy Saturday in July, she got a phone call from a friend. He had been hiking a trail near her home in Hollis, New Hampshire. He had spotted the hawk, looking drenched and obviously sick or injured, perched in a tree. But what could he do? "He had nothing with him to help him rescue the bird," she said—and besides, he was in a hurry. He was leaving on vacation the next day. Could Nan help?

"I was home," she explains to me. "So I grabbed a cat crate, a towel, and a pair of gardening gloves." She headed for the trail. In the parking lot, she met a girl near the trailhead. "Did you see a hawk on your walk?" she asked her. "I'm going to get him." The girl came along.

They hadn't walked far when they found the soaking-wet bird. "He's locked onto a branch and he's eight feet up in a tree," Nan tells me. "There was no blood or feathers on the ground," she says, "but I knew he was really sick. I knew I was his last chance."

Nan is strong, fit, and tough, but she is not eight feet tall. But she realized she could reach the bird if she stood on top of the cat carrier.

"I stood," she continues, "and looked him straight in the eye." He was cold, wet, and clearly weak. She had tried calling two other wildlife rehabilitators before she set out; she finally reached Maria, who told her what to do. She hesitantly threw the towel with her right hand onto the bird, as she'd been instructed. "And at that moment he came alive," she tells me, "flapping his wings but not moving from his locked perch." There was no time for Nan to be scared, she tells me: "I realized I had to do it—and right then!" With her left hand, protected only with a green gardening glove, Nan reached under the towel and plucked the hawk from the branch. She stepped off the cat crate with the bird in hand. Before she put the hawk inside the carrier, Nan asked the girl accompanying her to take a photo: the hawk's feet and legs are covered with a towel; one wing is extended; the bird's beak is open. Nancy looks radiant.

"It was a moment in my life I'll never forget," she tells me. "I did something I don't normally do. I wanted to make this happen."

She delivered the bird to Maria's rehab center, Wings of the Dawn, that afternoon. The next day, she phoned to see if the bird was still alive. "She ate a grouse for dinner, and two blue jays for breakfast," Maria reported. That made Nan happy; surely, she thought, that meant the bird was feeling better. But then, Maria told her "these birds eat till the moment of their death."

Maria let her know the hawk was a broad-wing, a youngster of indeterminate sex, and that it had been poisoned—probably having eaten prey that had been killed with toxins people unthinkingly set out to poison mice and rats. But under Maria's care, the hawk recovered completely. Nan kept in touch, and when Maria told her of the release, scheduled for September 19, she vowed to join her.

But on this day, I'm unaware of Nan's bravery or any of the details of the dramatic rescue as I clutch the bird's feet in one hand and cradle its wings with the other. I only know what the bird wants next. The Hawk Watch's organizers would prefer I release the raptor so it would fly over the gathered crowd, for the best photo; it's a small favor to ask of the bird, and a big reward for the people who support the counts and the research on birds of prey, especially when many species of raptors are declining at an alarming rate. But all I can feel is the bird's desire, burning as if it were my own: it wants to fly in the opposite direction, the direction in which all the other birds are flying, off the mountain, heading south.

I turn around, our backs to the cameras, swing my arms upward, and let go. The hawk flies free. Within thirty seconds, the bird is a speck in the sky—joining the twenty osprey, ten bald eagles, 105 sharp-shinned hawks, twelve merlins, two peregrines, forty-one kestrels, and 1,240 broad-wings that Hawk Watchers will have tallied on the migration by the end of the day.

⟳

I cannot imagine I'll ever see that bird again; even if I did, I wouldn't know it.

Certainly the bird has no reason to remember me, either. I don't matter to the hawk.

But the hawk matters deeply to me, and to Nan.

For her, rescuing the hawk "has opened a door I couldn't have pictured or planned. I look at it as symbolic in my life," she told me.

The last time she encountered such a powerful animal symbol, she was at a friend's cabin. She had never seen a Luna moth before—but suddenly, the huge, showy, pale green moths were everywhere. Moths and butterflies are widely understood as symbols of new life—transitioning as they do from wormlike caterpillars to pupae encased in their coffinlike cocoons, to flamboyantly beautiful, winged wonders. It was a time when Nancy was leaving an unhealthy relationship. She took it as a sign of transformation to come.

And what might the hawk symbolize in her life? I asked Nan. She doesn't know yet. "I think the meaning of the bird—or my energy to transition it back into this world, and not to death—is unfolding."

Since the release, we've both, independently and regularly, checked a website maintained by the Hawk Mountain Sanctuary, which tracks the travels of seven broad-winged hawks outfitted with either satellite or cell phone telemetry transmitters. The broad-wings were captured and tagged in Connecticut, Ontario, Pennsylvania, and New Hampshire. We have been particularly interested in the progress of an adult female named Monadnock.

Though it's tempting to imagine this is the bird we both briefly held, she is not; this one is an adult, not a youngster, and was captured and tagged on her breeding territory in nearby Dublin, New Hampshire, in June. Ours did not have telemetry. But we can hope, and reasonably believe, that the bird we held may have been traveling in her company, along with thousands of others. Before Thanksgiving, we saw with delight that she had already reached Colombia.

THE HAWK'S WAY

My introduction to falconry is a bloody one.

On a cool, gray mid-October day, master falconer Nancy Cowan, a petite, blue-eyed blonde in her late fifties, hands me the most beautiful bird in the world: a four-year-old female Harris's hawk (named after ornithologist Edward Harris) named Jazz. A deep, coppery brown, with reddish shoulders and a white tail tip, Jazz stands more than twenty inches tall, weighs thirty ounces, and her outstretched wings span nearly four feet. Her profile is regal and knowing. Harris's hawks are big, but their appetites are bigger. In their native cactus deserts in the Southwest, they hunt moorhens as big as ducks and jackrabbits that can weigh more than twice as much as they do.

Nancy has offered me a choice: of the dozen or so birds of prey she and her falconer husband keep on their rural New Hampshire property, I could work with Jazz or Emma, the lanner falcon. Emma is also beautiful. With a slate back and wings, creamy belly, and chestnut crown, her kind is the species pictured in Egyptian hieroglyphics. But Emma is smaller than Jazz by a third. And, Nancy explains, Emma has been raised by humans. At the mature age of sixteen or seventeen years old, Emma is exceptionally docile and calm.

Jazz is not. Nancy had warned me: Jazz is "feisty," sometimes uncooperative, and "doesn't like hats." (How Jazz expresses displeasure is left unspoken—but looking at her curved obsidian talons and sharp black beak, I am glad that despite the cold I have come bareheaded.)

But I am not afraid. It's her wildness I want from the moment I set eyes on her.

Stepping from Nancy's falconry glove to the one loaned to me, Jazz's huge yellow feet grip my left hand with shocking strength. It is wise to be sheathed in leather. Otherwise, simply by perching, her talons would rip the skin of my hand and wrist and could easily pierce me to the bone. I am impressed by Jazz's feet, but am awestruck by her huge mahogany eyes. They look past my face, past my soul, as impassive and hungry as fire. Her eyes seem to be devouring the world.

I know I don't matter to her at all. Yet to me, she is everything. Why do I love her so immediately? I love that she is big; I love that she is fierce; and I love, too, that she might be unpredictable. She is the essence of hawk, a bird so unlike anyone I have ever known. And here she is on my arm.

I pretend that claiming Jazz isn't greedy. That leaves the smaller, calmer bird, Emma, for my friend Selinda. It didn't take much to persuade Selinda to accompany me; she loves animals as well as a good adventure (her first job out of college was working as a geologist in Alaska). But attending the half-day introductory course at Nancy's New Hampshire School of Falconry was my idea, and I reckon if one of us might get hurt today, it should be me.

Our instructor chooses to work with Banshee, a teenage peregrine falcon. The bird of prey recently reintroduced to cities to help control pigeon and starling populations, peregrines dive through the sky after their prey at a heart-stopping two hundred miles an hour. Her head is capped in black, her back and wings a deep, shiny blue, like the skin on a blue shark. Five inches shorter than my big Jazz, she seems tiny, precise, a knight-errant clad not in chain mail but feathers.

We start walking down Nancy's gravel driveway, amazed that these majestic predatory birds sit sedately on our fists.

Then Banshee bites Nancy in the face.

The attack comes without warning. It's a hell of a bite. Later Nancy explains to us that being bitten by a peregrine feels like being punctured by a staple gun. A notch in the bird's curved bill, an adaptation for crunching the vertebrae of its prey, makes the bite particularly messy and painful.

Blood gushes from the wound. What will Jazz and Emma do? I worry they might attack at the sight of blood, as my chickens do, but they ignore it.

Selinda and I, however, gasp in distress. "Don't worry about it," says Nancy. "Banshee's a teenager. She's just being a brat." Though I periodically dab at the wound with a tissue, the blood flows down our instructor's cheek and drips off her chin for half an hour.

Nancy is used to this. And so apparently are the neighbors. Drivers slow their cars and wave genially. Only Selinda and I seem to think anything of the sight of three women walking down a country road with birds of prey perched inches from their faces, one of them dripping blood onto the road. Nancy and her falconer husband, Jim, have lived here for twelve years.

Everyone knows about their birds.

Only one vehicle stops. The driver greets Nancy (without a word about her wound) with the news "I've got something for you." He reaches into the back seat and pulls out a dead woodcock—a medium-size, sandpiper-like bird of New England's fields and meadows—and hands her the corpse through the window. Food for the hawks, I assume. But later, I learn this would be her dinner. Clearly, I have entered a strange new world. Selinda and I—vegetarians who mourn roadkills and weep over books in which animals are hurt—are taking this course because of our love of birds. We never thought that, less than thirty minutes into the course, we'd be facing violence, blood, and death. But what shocks me more is this: though I'm sorry that Nancy has been bitten and I'm distressed that the woodcock has been killed, somehow, in the presence of these birds, blood and death are not repulsive. I feel myself being drawn to a mind wholly unlike my own.

"God, birds are mean!" my friend Mike Meads once remarked to me. Mike had once been a farmer and had personally lopped the testicles off thousands of sheep; who was he to talk about meanness? Later, he had become a well-known New Zealand entomologist. I noted that insects were not known for practicing lovingkindness. But many others have remarked about this aspect of birds.

Even a pet parrot, who tells you he loves you in English, who showers with you each morning, and who shares your dinner at night, possesses a sort of ferocity. Living with a bird, as animal behaviorist Rebecca Fox told documentary filmmaker Mira Tweti, is not like living with a feathered child. "You should really think of it as living with an alien," she said. Even species of birds that have been kept as caged pets for centuries remain fundamentally wild.

Unlike domesticated mammals such as dogs and horses, who are generally grateful for food and affection, your bird may not like you. Your parrot, for instance, might get mad at you—for something like going on vacation or away to college—and stay mad for the rest of its long life. Your pet parrot won't hesitate to punish you, sometimes ruthlessly.

But this isn't meanness; it is something else. Some would call it savagery. I call it wildness.

Birds are like us in so many ways that sometimes we forget we are from widely separate lineages. Birds are wild in a way that we don't experience in our relationships with our fellow mammals. And nothing, I found, brings one closer to the pure wildness of birds than working with a hawk.

⟊

"People have such misconceptions about these birds," Nancy tells us. Surrounded by framed prints depicting scenes of hawking in medieval Europe, Selinda and I sit in the warmth of the tasteful dining room of Nancy and Jim's 1789 house for the lecture portion of the falconry course.

"Some people think, 'Nancy owns this bird, so it's a pet,'" Nancy says.

But Nancy's relationship to her birds is utterly different from that of a person to their pet. "They might live with me, I might feed them," she tells us, "but they are wild. These are predators. And that is the beauty of them."

Raptors' wild beauty feeds yet another popular misconception. Folks are drawn to falconry for all sorts of wrong reasons, Nancy says. And one of them is this: "People see the bird as an ornament—they are thinking how cool they will look with one on their arm."

Selinda with Emma the lanner falcon

One can see how a person might succumb to that allure. Falconry, "the sport of kings," connects its practitioner with a romantic, proud history, stretching back to ancient China, India, Egypt, Persia, and Babylon, thousands of years before the existence of Rome. At one time, the type of falcon an Englishman was allowed to own marked his rank: a king carried the gyrfalcon; an earl, the peregrine; a yeoman, the goshawk; the priest, the sparrow hawk; and a servant, the kestrel. It is a sport with its own battery of accoutrements, including beautifully tooled leather gloves for the falconer and elaborate, often feathered hoods for the birds, sometimes considered works of art.

Falconry also has a language all its own, known only by the shared brotherhood of fellow falconers. Some of the words are needed to describe the sport's many accessories: *jesses*, attached to leather anklets around the bird's legs, are soft leather loops to which one can hook a length of rope or a tether to the falconer's glove. The *bewit* is a slip of leather attaching bells to the feet, so you know where your falcon is. The *creance* is the long, light cord for tethering a falcon in training.

Special words describe the activities unique to training and caring for a bird of prey.

Imping is the act of mending a broken feather. *Manning* describes training the young bird to be carried on the fist. *Seeling* is

Banshee, Nancy's peregrine falcon.

Leather hoods, like these in Jim Cowan's extensive collection, keep hawks calm by shielding them from sights that might alarm them. Though many are also works of art, all parts of the hood are functional: the beautiful topknots, for instance, allow the falconer to seat it perfectly on the hawk's head as well as to remove it safely and comfortably.

K.C., Nancy's gyrfalcon, models a hood.

Red-shouldered hawks love tall woods and water. This one, photographed in Colorado, has an unusually light breast.

This red-shouldered hawk lives at the Blackland Prairie Raptor Center in Allen, Texas.

Peregrine Banshee enjoys a quail wing.

The peregrine's long toes and sharp talons seize other birds in flight.

Cooper's hawks are commonly seen in forests from southern Canada to northern Mexico, and today frequently inhabit suburban backyards and even cities.

The peregrine falcon is the world's fastest bird.

Photo © Tianne Strombeck

Scooter models his jesses and bells. Light, thin leather straps around the leg allow the falconer to restrain the bird on the glove while it is being trained; bells help the falconer locate the hawk in the field. The bells are attached by a light leather betwit so the metal never touches the bird's leg.

the word for the ancient, now-abandoned practice of sewing the bird's eyelids shut—temporarily deprived of sight, the bird is rendered dependent on the falconer and more easily trained.

But much of falconry's secret language underscores, like a promise repeated again and again, how special these birds are, how different from all other beings. Though many bird species hunt, kill, and eat other animals—from the shrike, a songbird also known as the butcherbird, who kills and then uses thorns to skewer other birds to store them prominently for a later meal and attract a mate, to the worm-eating robin—birds of prey are exclusively predatory. They are also known as raptors (daytime raptors, more accurately, to distinguish them from the unrelated and mostly night-loving but equally predatory owls). Sometimes they all are just called hawks. They include some three hundred species that go by various names: hawks, eagles, falcons, harriers, kestrels, kites. (And to make it more confusing, the English use different words than we do; for instance, their buzzards aren't our vultures, but what we would call our red-tailed and ferruginous hawks.) They live all over the world. They are the tigers of the air. They hunt like no other predator.

The language of falconry honors this difference. The falcon isn't sleeping, like ordinary birds or mammals; it's *jonking*. When it cleans its beak and feet after eating, it's *feaking*. The act of hiding the food with outspread wings and tail while it eats is called *mantling*. A bird of prey, in fact, is so rarefied that it doesn't even shit like the rest of us. Hawks "slice"; falcons "mute."

My interest in falconry, though, isn't in its language or tradition. From falconry I want only one thing: to get closer to birds of prey. Majestic, graceful, strong, big, brave, and smart: Who would not hunger for such company?

Of course, one can watch hawks at a distance—and I do. I'm always looking for them and frequently find them. They are more numerous than you think. Along any highway, you can sometimes count dozens, perched singly on trees near the road, looking for small animals among the mown grass. In the fall, from certain mountaintops, you can see hundreds, sometimes thousands, of normally solitary hawks in a day, migrating en masse to southern wintering grounds. In the summer, if you know where to look, the huge stick nests of certain birds of prey are easy to find. My husband and I are blessed with a pair of bald eagles who nest near a lake not far from our house. And one can also watch birds of prey in captivity, at zoos, nature centers, and attractions that offer flight demonstrations.

But I long for a more intimate connection, to see, close at hand, something of what it is like to be a bird of prey, to try to understand what is in their heads. As predators, what do they show us about other birds? I wanted to touch these birds' fine, ancient wildness, this pure savagery bereft of evil. And there is no other way to begin to do this except through falconry.

The writer Thomas McGuane calls falconry "one of man's oldest and most mysterious alliances in the natural world." It is the art of partnering with a bird of prey. And as Nancy says, "You are not the master. You are the junior partner." If your hawk doesn't consider you a good enough partner, the bird will fly away. At any time, a bird may leave the falconer—for days, or weeks, or forever. But I have also read of birds who so value their partnership with their human that they stick around even when they are not handled or confined. In Stephen Bodio's splendid *A Rage for Falcons*, he writes of hawks who don't travel to the hunting grounds confined in carriers, but fly into the car for the trip, fly out to hunt, and fly back in to go home. Steve knows a goshawk who lives loose in the woods but flies to the falconer's whistle to join him for a hunt. He knows a pair of wild goshawks who come to another falconer's hand.

Few people understand the true nature of falconry, Nancy tells us. Oddly, birdwatchers often look down on falconry; some consider it a form of slavery. Others dislike the birds themselves, for raptors not infrequently attack and kill birds at feeders. In the United Kingdom, rogue members of the pigeon-racing and game-shooting communities have been known to intentionally poison, trap, and shoot hawks, particularly peregrines, a crime that occurs so frequently that it could threaten the falcon's survival. In the United States, a fourteen-month investigation uncovered that thousands of peregrines and hawks in Oregon, Washington, and California were being killed by members of pigeon clubs who specialize in Birmingham rollers—a type of pigeon bred for a genetic anomaly that triggers a seizure midflight, sending the bird spiraling downward until it recovers before hitting the ground.

Naturally, no raptor can resist such an easy and obvious target. But the breeders who perpetuate the defect persecute the predators by trapping, shooting, poisoning, clubbing, and, in some cases, torturing the raptors, including dousing the birds with Drano. Perpetrators convicted in 2009 received no jail time, prompting public demand for tougher penalties. Though today poachers can face six months in jail and a $5,000 fine for each illegally killed raptor, when a California man was convicted, in 2019, of killing a record number of birds of prey—a mass murder of more than one hundred raptors—he served only ninety days in jail. He was fined $75,000, also way below the maximum penalty.

But even those who admire hawks and falconry often misunderstand them. "Some people come to me with a mystical outlook," Nancy says. "They think, 'Oh, this bird loves me.' They think they have some kind of spiritual relationship with the bird." This view, she tells us, is as misguided as considering the hawk a pet or an ornament—and in its different way, just as demeaning. "I can't imagine anything crueler to do to a living being," she says, "than to try to make it into something it's not.

"The only people who understand birds of prey," she says, "are people who have worked with them for a long time. And every time I work with them I learn something new. You accept that your life will be changed by being a falconer."

Though I do not know this yet, Nancy is right: My life will change, too. Over the course of our long, continuing association, Nancy and her birds will show me a kind of relationship I had never known was possible with any living being.

The saker falcon in the hallway looks like a sculpture. The bird sits on its perch immobile, its head covered with a leather hood. Hoods have replaced the old practice of seeling the eyelids of a hawk in training. You would think that being blinded would send a bird into a panic, but no; a hooded bird doesn't thrash or struggle; it won't bite you or grab you with its talons. Instinctively, the bird knows better. It stays so calm that you might not even know that a hooded bird is alive. And in a sense, it is not. Putting a hood on a bird is like extinguishing a candle.

But then Nancy performs a breathtaking act of magic. She pulls off the hood. Instantly, the inert sculpture comes vividly alive. The flame of his soul leaps to life. Brown with a pale head, he stands tall and alert as an officer at attention: His name is Sabretache, for an accoutrement of the British military uniform, a flat pouch that hangs from the saber belt. The intensity of his gaze fills the room.

"These eyes are carrying huge messages," Nancy tells us.

One of the defining characteristics of birds is the crucial role and astonishing acuity of their vision. Flight, after all, demands excellent eyesight. For birds that hunt on the wing, the eyes are developed to an extraordinary degree: an eagle riding a thermal at one thousand feet can spot prey across a distance of nearly three square miles. In this way, raptors are superbirds. They have developed the avian sense of sight to perfection. A raptor's vision is the sharpest of all living creatures.

All birds' eyes are huge in proportion to their bodies. A person's eyes take up only two percent of the face; a European starling's account for fifteen. A great horned owl's eyes are so enormous relative to its head that if human eyes were comparable, they would be the size of oranges. Their eyes are so important to birds that, like various reptiles, sharks, and amphibians, they have a transparent or translucent third eyelid, the nictitating membrane, to protect and moisten the eyes while retaining visibility. Vision literally sculpts birds' every movement: one reason birds seem to move in such a jerky manner, as cassowary expert Andy Mack explained to me, is that the bird is actually keeping its head remarkably still, thanks to an extremely supple neck, while the rest of the body is in motion, in order to allow it to focus on what it sees in exceptional detail.

In birds of prey, the eyes weigh more than the brain. The

two eyes are twice as large as the brain itself. They need to be huge. They are packed with receptors, some types of which humans don't have at all. Raptors have not merely two (as we do) but six types of photoreceptors in the eye. Because of this, birds are thought to be able to experience colors that humans cannot even describe. Their retinas, unlike ours, contain few blood vessels. Instead, a thin, folded tissue called pecten, unique to birds, brings blood and nutrients to the eye without casting shadows or scattering light in the eye as blood vessels do.

Most birds, like most mammals, have a single area within the eye of perfect vision, called the fovea, where cone cells, which detect sharp contrast and detail, are most concentrated. A raptor's eye has two foveae. One is for lateral vision, the other for forward vision. A human eye has two hundred thousand cones to each square millimeter of fovea. Sparrows have twice that. Raptors have about a million.

Raptors see in such fine detail that humans need microscopes to begin to imagine it. They also have a wider field of vision than we do, thanks to the second fovea, as well as better distance perception than other birds. Most birds' eyes lie at the sides of the head so that when they look at something, they use one eye at a time. With forward-facing eyes, raptors have binocular vision like ours, but better. Fields of view of the left and right eye overlap, allowing the brain to compare the slightly different images from each and instantly calculate distance.

And there is something else about a raptor's vision, something more difficult to describe. "These birds don't think the way we think," Nancy tells us. "They don't learn the same way we do." Because of our differing brain circuitry, birds capture at a glance what it might take a human many seconds to apprehend. For all birds, but especially these, seeing is not merely believing; seeing is knowing. Seeing is being. That is what I see in Jazz's monstrous, devouring eyes: the windows to a mind completely different from my own.

"It's instinctive," says Nancy. "It's not spiritual. A falcon is at once the stupidest thing you'll ever deal with—and the most instinctively developed thing you'll ever deal with."

Instinct gets short shrift among most humans. We value thinking instead and dismiss instinct as the machinery of an automaton. But instinct is what lets us love life's juicy essence: instinct is why we enjoy food and drink and sex.

Thinking can get in the way of living. Too often we see through our brains, not through our eyes. This is such a common human failing that we joke about the absentminded professor or the artist so focused on his imagined canvas that he walks into a tree.

But Jazz won't smack into a tree. We are out in the field

across the street now, and Nancy unclips the jesses that keep Jazz tethered to my glove. "Let her fly," says Nancy. I give Jazz a brief toss from my glove, and she sails into a pine. She looks down at us. Now I am worthy of Jazz's interest. She knows something is about to happen. For the first time, I am bathed in her sight. It's a baptism, and feels momentous, transforming.

"Now call her in," says Nancy. She takes a piece of cut-up partridge out of a baggie in her pocket and places it between the thumb and forefinger of my glove. "Jazz!" she calls. I extend my left arm and look up. A huge, powerful bird flies toward me.

Not everyone would like this, I realize. An exceptionally brave biologist with whom I have worked in Southeast Asia, hiking in search of bears among forests littered with unexploded ordnance, confesses he would be scared. It's a genetically pro-grammed human reaction. Birds like this once hunted and killed our ancestors. A famous fossil hominid, the so-called Taung Child discovered in South Africa in 1924 and described by Ray-mond Dart, bears the marks of this predation. When I was in college, we had been taught that this long-dead australopithe-cine child must have been killed by an ancient leopard. Now, from careful reexamination of the skull, we know that the death blow dealt to the brain came from the talons of an ancient rel-ative of the crowned hawk-eagle—a raptor that still hunts large monkeys in the same manner today. Our kind has rightly viewed birds like Jazz with caution for more than two million years. No wonder so many people flinch.

But as Jazz's talons reach for my glove, my heart sings.

She lands surprisingly heavily for a bird. The squeeze of her talons is strong. Her piercing eyes are focused on the job at hand: tearing flesh with her beak and feet with an intensity that encompasses at once rage and joy. Though Jazz has done nothing more than land on my hand, I feel she has given me a great gift.

Sy with the Harris's hawk Jazz

On my hand, I hold a waterfall, an eclipse, a lightning storm. No, more than that. Jazz is wildness itself, vividly, almost blindingly alive in a way we humans may never experience.

This is one reason I have always been drawn to animals: their sharpened senses give them a fuller experience of the world. Largely oblivious to the symphony of scents, humans experience only a small part of life. We hear but a sliver of the range of the world's voices and have evolved to depend on vision most of all. But although we live through our eyes, birds do so to an even greater degree.

Birds' eyes gather more of life than ours do. Perhaps this is why I could feel Jazz so purely, densely full of life, filling up the moment—here, now, and nothing else. The Buddhists say there really *is* nothing else, because now is timeless; now is everything. Perhaps because of this, Jazz seems more immediately alive than any human I have ever known. To be in the gripping gaze of that bird is like looking directly into the sun. The class is a transforming experience. I am hungry for more.

⟨∾⟩

Two years later, I never forgot Jazz. I longed for the laser intensity of her eyes, the assured passion of her instinct. I longed to be with her and know more about her kind. But life intervened. A pressing book deadline; an expedition to New Guinea; a new hatch of baby chicks; a rescued border collie; two national book tours; an expedition to Mongolia.

But I knew I'd return. And now I am back in Nancy's New Hampshire farmhouse kitchen, with its green checkered curtains and worn wooden countertops on which endless piecrusts have been rolled. With the cheerful air of someone baking cookies for her grandchildren, Nancy is using a bread knife to cut off the heads and legs of frozen baby chicks and popping out the hearts to use in training the new Harris's hawks.

"They love the heart," Nancy is telling me. "They also love the heads."

I try not to think the meat she's cutting was once sweet baby chicks, like the ones I've been raising each spring at home for decades—fluffy infants who sleep in my sweater and perch peeping on my shoulders and head. I am trying to stay focused on why I came here.

I came back for Jazz. Or so I thought. When I phoned Nancy to set up an appointment, I learned to my shock and sorrow that Jazz was dead. The majestic hawk had never shown a sign of weakness; her appetite was sharp, her flight strong, her feathers perfect. But one day, a year before I called, when Nancy went to visit Jazz in her aviary, the bird looked sick, and when she picked her up, Jazz died in her hands. A necropsy showed it was cancer.

I was so stunned and sad I couldn't speak. I was loath to let on to Nancy that Jazz had meant so much to me after our single half-day encounter. I knew I could not have meant anything to Jazz; I doubted that Jazz would remember me and wondered whether even Nancy would. But she did.

"Before you came, two years ago," Nancy said gently, "I wasn't sure Jazz would fly to anyone but me." Previous attempts had failed. She just wouldn't land on anyone's glove but Nancy's—until mine.

Jazz had, in fact, given me a great gift. I would go forward with my study of this discipline, I decided right then, in her honor.

"Some people take to falconry easily, and others will never be a candidate for this," Nancy continued. "You need to read the bird. For some that comes hard. For some it comes easy. And frankly, Sy, it comes easily to you. You are working in the same time frame, flowing with what the bird is doing."

This is what keeps me from retching as Nancy cuts up the chicks and pops out their hearts. To flow with a hawk, to enter its timelessness: I want this so bad I can taste it.

〜

Nancy and I have just come in from greeting the new crew. On this fine, cool October morning, I will be working with young

Harris's hawks that hatched at a breeder's this summer. They're all just about as big as adults, and impressive birds. But one look in their eyes shows they're still just babies: their eyes aren't mahogany like Jazz's, but still a gray blue.

In the side yard are their aviaries, called mews. Each bird has its own separate red wooden building about twice the size of a spacious toolshed. Next door to Sabretache, the saker falcon, lives Smoke, the larger of the two Harris's hawk sisters. Long before Nancy even thought of teaching Smoke to chase a flying lure to prepare her for hunting, Smoke was chasing after leaves that flew off the trees. "She's going to be a wonderful hunter," says Nancy. Though Smoke is seven ounces heavier than her sister, her plumage looks less like that of an adult. Unlike Jazz's rich brown, Smoke's breast feathers are the color of mocha ice cream drizzled with melted chocolate.

Smoke's smaller, darker sister, Fire, lives in the mews across from her—next door to Moseby, a female goshawk, and two doors down from Banshee (who, Nancy tells me, has overcome her crabby adolescent stage and become a fine companion hunter). Fire is screaming bloody murder. Nancy calls the racket "peeping." Baby hawks do this when they want attention from the parents. "Fire's smart," Nancy tells me, "but she is stunted in her emotional development." Her sister used to steal her food when they were housed together at the breeder's, and this made Fire babyish and demanding.

Out back are more birds. Nancy only introduces me to two Harris's hawk brothers who hatched from different clutches but look identical. Scout hatched in May and is eight weeks older than Sidekick. But Sidekick is the fearless one. That's sometimes a problem. "This guy lands on my head, on my shoulder—and he talons me! He opens me up. Jazz used to grab me like that," Nancy remembers. "But he's a smart cookie. He'll learn."

Harris's hawks are smarter than other hawks, Nancy insists. "A goshawk will watch a rabbit go down a hole and just sit there," Nancy said. "A Harris's hawk will realize your dogs will chase the rabbit out of the hole, and he'll go wait for the rabbit to emerge on the other side." Harris's hawks are naturally companionable. Sometimes a female (as with all raptors, the larger sex) will take two mates at once. Like crows' young, those from a first clutch of Harris's hawks often stick around the parents to help them feed the second clutch. And unlike most other birds of prey, in the wild, Harris's hawks often hunt in packs. Cooperative groups of up to a dozen may gather to hunt hares, birds, and lizards in the Southwest desert, especially in winter. Harris's hawks are born genetically programmed for partnership, about which I am just beginning to learn.

"Doing is the best way to learn," Nancy tells me. Traditionally, she says, an apprentice learned by mostly just watching a more knowledgeable falconer. But today I'll be able to do more than just watch; with four young birds to train, she's giving me the great compliment of allowing me to help. "You'll fly a bird today," she tells me.

We enter Fire's mews. Nancy pushes her glove up to Fire's breast, and the bird steps onto her glove. Inches from Fire's ebony talons, Nancy places a clip through a hole in the jesses to tether the bird to a rope tied to the glove. But as soon as we leave the mews, Fire tries to fly. She launches—but because she is hooked to the tether, she flips over. She hangs upside down from the leather straps on her great yellow feet, thrashing the air with her huge, powerful wings.

Nancy reaches toward her with her naked right hand and gently pushes on Fire's back to swing her back up to the glove—impressively without tangling the jesses, being bitten, or getting taloned. But Fire flies and flips again. Nancy again scoops her back to the glove.

This strange behavior is called *bating*. Why would a hawk want to hang upside down from your glove by its feet? "They don't want to hang upside down," Nancy says. "They want to fly." Why don't they figure out that if they try to fly while they're tethered, they're going to hang upside down? "Well, they just don't." These are such strange creatures. How will I ever begin to understand what is going on in their heads?

Nancy has to weigh Fire to make sure she's not too full from her last meal to want to follow me. She urges the bird to step onto a scale. Fire weighs twenty-six ounces and is hungry. Nancy unclips the line tying her to the glove and lets her fly onto the branch of a pine in the side yard. Nancy tells me to walk away with my back to Fire, and then turn around to see if Fire is watching me.

She is. I can feel Fire's gaze even before I turn.

I take perhaps twenty steps before Nancy tells me to call the bird. Surreptitiously I reach my ungloved right hand into the pocket of my jacket to part the lips of the baggie hidden there. My fingers find the cold, bloody slime of a cut-up chick. I feel for the point of a beak; I want to offer the coveted head. I place it between the thumb and forefinger of my borrowed glove. I turn toward Fire and extend my arm.

I desperately want to do everything right and remember all that Nancy's told me.

"Present the back of your shoulder to her," Nancy had said. I turn so that my back is to the hawk, extend my gloved hand, and look at the hawk over my left shoulder. This is a stance that could protect me in case I ever fly a goshawk. Goshawks not infrequently attack your eyes, which is how many hawks naturally kill their prey, forcing their talons through the skull via the eye socket into the brain, Nancy explained. That's why "if they think there is any chance the food might go away, they fly into your face and go for your eyes."

Nancy learned this firsthand with a goshawk who belonged to her husband, Jim, from whom she learned falconry when she was his apprentice. It was their wedding anniversary, and Nancy was all dressed up waiting for Jim to get home from Boston to go out to dinner. Arriving late, he asked if she would feed his new goshawk while he showered and dressed. It would be easy, he told her: "Just present the first chick on the glove. Let her eat it. Present the second chick. Let her eat. Present the third. Then you're done."

"So," Nancy told me, "I go into the mews, she's sitting there. I present the chick. She eats it. No problem. I do the second one. Fine. I give her the third and I turned sideways to leave—and I felt talons going around my eyeball!" She had two puncture wounds on her face, both bleeding copiously. "If she had clasped her talons," Nancy says, "she would have torn my eyeball out."

Jim found his wife leaning up against the door of the mews, bleeding and furious—not at the bird, but at her husband.

"What didn't you tell me?" Nancy demanded.

"I told you to present three chicks . . ."

"What didn't you tell me?!"

And then Jim remembered. "Oh, I forgot—after the third chick, I reach out and go *fluffel, fluffel, fluffel* on her breast feathers."

Just that one change in routine was enough to provoke an attack. "That's how much of a raw nerve they are," Nancy said. If things aren't going as expected, the best plan is to attack. "They're acting on an instinct that is so fast, it's like lightning in a bottle."

That's why she trains her students to present the back of the shoulder to any incoming hawk, she told me. "Should you ever fly a gos, and it decides to go for your eyes, you can turn and shield your face with your ungloved hand."

This is what you are dealing with when you work with a bird of prey. Though most species are far less aggressive than the goshawk, all of them can be unpredictable and dangerous.

"It's like handling a loaded gun," my husband, Howard, later said with alarm.

That lesson is not difficult to remember. But I am not thinking of this as Fire flies to me, talons reaching for my glove. I am so hungry to have her land on my fist, there is no room in my heart for anything else. My whole soul feels like a yawning hole that only this bird can fill. She lands on my glove—*smack!*—and begins tearing at the chick head. I pull my arm in to draw her closer to me.

Were I to find a crispy drumstick on my dinner plate, I would feel ill. But as she eats the meat, her joy is mine.

What is happening to me?

∽

Weeks progress. Thanks to Nancy's clarity and patience, slowly I am beginning to know how to move, how to watch, how to think around hawks.

At first, it seems like a lot to keep in mind. As in all new endeavors, there are strange little injunctions to remember—like never say *Shhh!* to a hawk. (It's similar to the warning hiss they give before biting.) There's a lot of equipment to learn how to use—never my strong suit. There is a certain amount of technique—how to hold the bird (always higher than the rest of the arm), how to "load" the glove (without letting the hawk see you remove the bait from your pocket), how to attach the tether to the jesses (quickly, before the bird "foots" you—the falconer's term for when a bird tears its talons through your flesh).

Getting things right really matters. If you make a mistake, you can be hurt. But far worse, your bird can die. A hawk's wings, tail, feet, and eyes are so delicate they can be easily injured when you take the bird out of the mews. Even an unexpected wind can lift, twist, and sprain a wing when a bird is tethered to your glove. You must be careful, especially with Harris's hawks, who are native to warmer climes, not to fly a bird on a cold day. Even mild frostbite on the feet can kill. Infections can be lethal. Tangled jesses can break a leg. The list of hazards goes on and on.

Because they have regular food and access to veterinary care, hawks used in falconry generally live much longer than those that are free: eighty percent of wild hawks die in their first year; a falconer's might live to thirty. But even with human care, a hawk's life is fraught with danger. One of the greatest heartaches a falconer can know could happen each time you take your bird out of its mews: a wild hawk in the sky might attack and kill your bird.

Nancy knows this sorrow.

She was out hunting with a young peregrine falcon named Witch. Nancy had raised her from a downy, helpless chick. The name was short for Kitchen Witch, because during her chickhood, during the day, the kitchen was where Witch stayed, to be near Nancy all the time. At night Witch slept by Nancy and Jim's bed. Young raptors are irresistibly ugly babies—fat-bottomed tripods of fluff with unnaturally huge feet that chase Coke cans across the floor, scream at you with eyes full of wonder, and then fall on the floor, facedown, legs out, asleep. To raise a bird from this endearing, silly-looking chick to a powerful hunter demands a huge investment of love and work. And now Nancy and Witch were out hunting, cementing their partnership.

Witch flew over an orchard after game—and flew back with a red-tailed hawk in pursuit. Again and again the big redtail hit the young peregrine. Witch landed in a tree, and though Nancy tried to call her to the glove, where she would be safe, Witch was too young and afraid to chance flying low—the higher hawk always has the advantage. Nancy raced off to try to get her car, parked far away, to get closer to Witch. She stood in the street and flagged down a logging truck to hitch a ride. Within twenty minutes she was close to the last tree in which Witch had landed. There, on a low branch, sat the redtail. It looked much fatter than before.

For five days, Nancy returned to the spot, calling for Witch, swinging a lure baited with meat till her hands were bloody. But she knew what had happened. The redtail had eaten her bird.

Once, as she was cutting up baby quail for the hawks, Nancy told me her motto, one she credits to a fellow hunter, Teddy Moritz: "Hunt hard. Kill swiftly. Waste nothing. Offer no apologies."

"It's nature's way," she says. And when Witch was killed, that, too, was nature's way. But Nancy still mourns.

∽

Slowly, I am starting—just starting—to understand who birds of prey are.

Sometimes they seem shockingly mindless. By my fourth lesson, it's clear Fire knows me. She doesn't move away from me when I approach her perch. She eagerly steps onto my glove. Yet as I carry her toward the door to go outside, almost invariably she launches off the glove and hangs upside down, bating. She can't possibly enjoy this—yet she does it again and again. Why can't she seem to remember the consequences and just wait till I unhook her?

Because, Nancy explains, so much of hawks' behavior is hardwired, instinctive. Fire bates off the glove because the urge to fly is overwhelming.

Equally overwhelming is the reaction to clench. A hawk who has caught a rabbit may not let go until the rabbit dies—or until the hawk does. If a rabbit races into its hole with a hawk on its back, the hawk will allow itself to be taken underground—and rather than let go, it will die there. Why? "It can't let go," Nancy explained to me. "If something wiggles, they bind their feet to it as a reflex. They don't have the choice. The mind is not involved in this at all."

If you are footed by a hawk, you must remember this. One of Nancy's apprentices, Rita Tulloh, discovered this when her huge red-tail hen, Scarlett, nailed her hand. Your instinct is of course to pull away, with the bird's talons deep in your flesh, but you mustn't struggle, or else you will never extract yourself from its grip. "The clench doesn't even have to go up through the thinking part of the bird's brain," Rita told me. "It's like a sneeze or a blink. Meanwhile, you just wait and drip." Rita will always have a scar.

But in other ways, a hawk's mind is like a steel trap.

"They have the instinct to hunt and fly," Nancy told me, "but *how* to hunt, they learn. It's as if there's some file folder in their heads about hunting success that they learn and never forget."

When it comes to hunting success, hawks learn instantly—and once they do, they remember it forever. Nancy told me about the first time Jazz went hunting. She followed a pheasant sixty feet up in the air, grabbed it, caught it—and then had no idea what to do. She let go and was frustrated and mystified when it disappeared.

"But she never let that happen again," said Nancy. The next time she grabbed a pheasant, Jazz held on and killed it. When hunting, a hawk doesn't make the same mistake twice.

They are not automatons, however, and can be quite emotional. Nancy told me about the winter day she was hunting with her first Harris's hawk, Indy. She slipped on the ice on a small bridge and fell into a shallow stream. Indy was unhurt, but he was furious: "He thought I had thrown him down!" she said. He screamed insults into her ear and remained angry with her for a week.

Anger and frustration tend to be the emotions they most often show. (Oh, great—a loaded gun that's mad at you.) At least this is one emotion that's easy to notice: I had no trouble recognizing Scout's disapproval when he saw me wearing a black headband for the first time. "WRAAACK! WRAAAC-CKKK! WRAAACCKKKK!" "WHO THE HELL ARE YOU? GO AWAY! GO AWAY!" He screamed at me like he wanted to kill me. He probably did want to.

I am learning how to read these birds' postures and gestures, understand when they feel comfortable and when they do not. I know when Fire is going to bate off my glove and how to deal with it. I remember that Scout doesn't like it when I present my shoulder, so I present my chest (remembering not to do this if I ever fly a goshawk). I know that Smoke and Fire both recognize me by now and are not annoyed by my presence. I feel confident handling their jesses and even touching their feet. If I can learn how to behave correctly, perhaps these gorgeous creatures—whom I love like an Aztec loves the sun—will learn to trust me not to do something scary or stupid.

One day, before an event at which Nancy would fly Fire before a crowd, I held Fire while Nancy gave her dirty tail a bath in the kitchen sink. I held her jesses firmly between the thumb and forefinger of my glove, expecting she would bate or even try to bite or foot me. To my surprise, she was calm, and I felt a wave of tenderness sweep over me for this powerful predator during a lovely, intimate, and quiet moment. I think she rather enjoyed her bath.

But I know not to look for affection from these birds—now or ever. Falconers, each of them crazy in love with their birds, accept that their birds will never love them back. At least not in the way a fellow mammal shows you its love. From these birds, I want something else—if I can earn it.

"They may show you a certain companionship. They can become comfortable with you. With Indy," said Nancy, "we knew each other so thoroughly, I could close my eyes and know exactly where he'd place himself in a tree relative to me. It's a beautiful partnership. But if you break their laws, you'll pay."

"If you want love out of this," Nancy tells her students, "you're too needy. Don't be a falconer."

For a human to love without expecting love in return is hugely liberating. To leave the self out of love is like escaping the grip of gravity. It is to grow wings. It opens up the sky.

∽

Nancy is teaching me how to work with the lure—a tool to help young hawks learn to chase and catch prey on the wing. The lures with which we work are leather, one in the shape of a rabbit, the other like a bird. They're baited with meat. Thrown to the ground or swung on a rope in the air, the lure inspires pursuit and helps the young hawk practice maneuvers to attack and seize its prey.

Nancy takes out Scout to show me how it's done. She swings the flying lure, baited with a chick torso. He makes a pass but misses, lands in a pine tree, and watches. She swings it again. He misses twice more. He screams. He seems to be tiring, nearing frustration. He wants to catch it but hasn't the skill. The fourth time, Nancy swings the lure as slowly as possible—and he seizes it in the air and tackles it to the ground. He spreads his wings and tail over the prize, mantling, as he tears at the meat. It is an act so intimate, private, and selfish it feels as if I ought to avert my eyes. But of course I cannot take my eyes off the bird. Nor do I really want to.

Nancy shows me how to retrieve the lure. While the bird is eating, you step behind the bird, onto the lure, taking care not to hurt the tail feathers; you load your glove with bait, but keep it hidden; and when the bird is finished eating, show it the meat on the glove. The second the bird flies from the lure to your glove—again, while it is occupied with eating—turn your body to retrieve and hide the lure. Nancy stuffs it in a back pocket of her vest and attaches the bird's jesses to the leash.

It looks simple. Now I'll try it with Fire.

The bunny-shaped lure is baited with a chick wing. I toss it on the driveway. Fire flies to it and begins mantling as she eats it. I am lost in her pleasure, lost in her beauty, drowning in my love of this bird.

I forget to load my glove. Nancy reminds me. And then I forget to step on the lure. Quickly Fire finishes eating, and unwisely I show her my baited glove. Up she flies to my fist—her talons still firmly grasping the lure. "Oh, that's bad," says Nancy. Indeed, it is—now it will be difficult to detach her feet from the lure. Nancy has me put the bird, glove, and lure back on the ground—now I am thoroughly flustered—and again shows me how to do it right. Fire now flies to my glove and leaves the lure behind.

Later we discuss why letting Fire bring the lure to the glove was such a big mistake. Between the string to the lure and the string to the glove, Fire could get dangerously tangled. Trying to escape from the tangle would only cause her to tighten her grip on it more. "Plus," I said, thinking I could apply a lesson from dog obedience, "allowing her to take the lure to the glove and getting two bits of chicken rewards her for a behavior you don't want to reinforce."

Nancy corrected me emphatically. "You must never think of rewards or punishments," she told me. "If you think in terms of rewards and punishments, you're not thinking partnership. They don't serve us. We serve them."

In falconry, you don't train a hawk to do things for you. In *A Rage for Falcons*, Steve Bodio puts it this way: You train a hawk "to accept you as her servant."

"We are not giving them anything," Nancy stressed. "It's either theirs or it's not. We are working with the bird. It's a partnership. And even after all these years, I'm the junior partner. I am not the master."

This is the biggest mistake I have made so far, and I am feeling exceptionally clumsy. My hands won't work in the cold. My glasses, which I wear for distance, blur close-up work, like linking the French clip to the hole in the birds' jesses. But as the day's lesson is about to end, Nancy surprises me with a new assignment.

"I want you to invest in a whistle. To call the bird. The whistle means food. It will make you more independent." That will mean stopping by the hunting goods store—how odd, I think: me in a hunting goods store.

"And I want you to order your own falconry glove." It must be leather—nothing else will protect against those talons. I never buy leather, but eagerly I find myself taking down the number for the falconry supply outfitter.

"You'll have the whistle," Nancy continued. "You'll have the glove. There's only one thing missing."

"What's that?"

"The bird."

⌒∾

Nancy is inviting me to be her apprentice. This is a huge honor, and an act of great generosity on her part. It means Nancy can no longer collect the checks I write out for lessons. If I am her apprentice, she teaches me for free.

But the apprentice makes the larger investment. It's a two-year commitment—and usually leads to a lifelong passion.

First you must complete a weekend hunter's safety course. Pass the exam for a hunting license. Build a mews and set it up with all the furniture a hawk would need. And then, sometime between September 1 and November 30, catch a young wild redtail and make it your own.

Could I do this?

There are several ways to catch a wild hawk. One way is to prepare a trap called a bal-chatri. This ancient device, invented in India, looks like a loosely woven upside-down basket, covered with small nooses. You can make a bal-chatri from slats of cane, hardware cloth, or kitchen wire—for a small bird like a kestrel, even a kitchen strainer will do. Beneath the trap you place a live rodent or pigeon to attract your quarry. Place it in an area you have seen young birds frequent, and go off and hide in the bushes or your car. When the hawk lands on the bal-chatri and tries (unsuccessfully) to grab the bait, its feet tangle in the nooses. You then rush out from your hiding place, throw a towel over the hawk, disentangle the feet, and take your captive home.

Capturing the bird isn't always easy, but that's not what worries me. My concern is this: Could I in good conscience take a bird out of the wild?

Other committed conservationists do. After all, a falconer's bird will almost certainly live longer than a wild one, and once your apprenticeship is over, you can set it free. Having made it through that first, most perilous year, the hawk will have an even better chance of survival than if it never met you.

But what about my chickens? They free-range all over our property. How could I keep a hawk on the property with chickens, too?

Even if I managed to work with the hawk away from the chickens, even if I set her free after a year, wouldn't she find me? I imagine looking up into the sky to see my hunting partner coming back to me, wild and free—and then slamming to earth like a lawn dart onto the body of Pickles or Matilda or Peanut.

Nancy and her hunting partner, Harris's hawk Scooter.

Scooter flies eagerly to the glove.

Nancy hunting with Scooter and her Gordon setter, Rain.

About the size of a mourning dove, the American kestrel is North America's littlest falcon. You'll often see them perched on telephone wires. This one was photographed at Blackland Prairie Raptor Center.

A fish specialist, the black-collared hawk hunts in wet lowlands throughout much of the Americas, especially in the tropics. This one was photographed in the Pantanal, the world's largest tropical wetland, in Brazil.

The most common hawk in North America, a red-tailed hawk, like this one, is easy to spot perched on trees beside roads or wheeling in flight over open fields. Its distinctive two- to three-second scream is usually uttered while soaring, warning other hawks off its territory or telling humans to back away from a nest.

The rough-legged hawk nests in the arctic tundra, hunting mainly lemmings and voles. Tied to cold climes, in winter they seldom venture farther south than the central United States.

The impressive great black hawk loves to hunt near lakes, ponds, streams, swamps, rivers, and marshes throughout most of South America and through to northern Mexico. Though it can soar quite high, it sometimes hunts by walking on the ground. It's known to seek out grass fires to capture small animals fleeing from flames and smoke.

One glimpse of the chickens, and they would forever be filed under "hunting" in the hawk's steel-trap mind—causing her to return again and again, until every one of our beloved, gentle, harmless hens had been killed, pierced through the brain by her cruel, perfect talons.

I ask Nancy.

"The hawk would kill the chickens," she tells me flatly.

I pose the question that night to friends over pizza at a restaurant. Could I be a falconer? One of them is the author-illustrator Lita Judge. Her grandmother, Fran Hamerstrom, a legend among falconers and birders alike, was a pioneer in banding hawks and rehabilitating injured birds, and the first to successfully breed endangered golden eagles in captivity. Lita grew up with hawks and owls swooping through the rooms of her grandmother's house and knew how to handle them all. What does she think?

Lita hasn't gone near hawks or owls for years, she says. She has purposely avoided working with them. "I'm afraid it would take over my life." Falconry, she warns, is addictive.

"One thing you might not want in your life," she says, "is lots of dead things in your house all the time." My husband puts down his pizza and rolls his eyes.

Lita looks at my hands. "Also, the skin on your hands is pretty thin. There's not much flesh there. A hawk could give you a pretty disabling injury. If you got footed, it would go right to the bones and the ligaments and tendons." Some eagles used in falconry, the golden eagles of Mongolia, can crush and break the bones in a human hand with a slight flex of their talons. "That's a consideration for you," Lita says, "since you take notes with your right hand."

But what about the chickens?

Lita doesn't even have to think about that one. "The hawk would kill the chickens, all right. No question about that."

Could I find some way around this? Could I keep the hawk somewhere else, off our property? My friend Liz Thomas, who lives seven miles away, generously offered her home, which is inhabited by two Australian cattle dogs, two parrots, several cats, and one husband—but no hens.

But then there is the issue of my travels. Besides short trips for speaking engagements and book tours, I am often away for a month or more at a time working in some remote jungle or desert. In my absence, my husband and neighbors have, over the years, cared for a menagerie of critters ranging from our 750-pound pig to hens, ferrets, dogs, and parrots. But in good conscience, can I ask another person to tend a loaded gun?

As I tell friends about learning falconry, many are excited for me, but as I describe it further, some grow visibly distressed. One, a biologist, now eighty, who flew a sparrow hawk at thirteen and likes to tease me about my vegetarianism, asks with concern: "*You?* A hunter?"

Elizabeth, our tenant, who shares our house, and whose own flock of chickens occupy one bay of the barn, is too kind to say anything judgmental, but I can see it on her face. She hates hunters.

Many people, knowing how I love animals and eschew meat, assume that I hate hunters, too. But my mother was a hunter. Growing up in rural Arkansas, she hunted squirrels and ate them. My friend Liz, who had offered her home to me should I get a hawk, is also a licensed hunter—though she hasn't actually shot a deer. Other friends are hunters, too, many of them tireless conservationists, people I deeply admire. Hunting is a far more humane and ecologically sane way of obtaining meat than factory farming, which is hideously cruel, wasteful, and, because it is a leading cause of climate change—not to mention that it generates tons of concentrated animal waste—ecologically disastrous. I understand hunting for food. But unlike a cheetah, a polar bear, or a falcon, I don't need meat to survive. So, rather than taking up hunting, I gave up meat.

But falconry, I realize, is hunting. Venery. For years, as a child, I had believed that venery was one of the seven deadly sins, like envy, greed, and sloth. But no, *Webster's* assures me I was wrong. *Venery* is defined as "the art, act, or practice of hunting."

A second meaning, however, is listed in the *American Heritage Dictionary.* "Venery: the indulgence in or pursuit of sexual pleasure." The word is rooted in the Roman goddess Venus, whose name meant desire or love.

"Gaia put the will to hunt deep in our psyches," my friend Liz, whose family was the first to study the Bushmen, now called the San, in the Kalahari, observes in one of her many books on people and animals, *The Hidden Life of Deer.* We have hunted throughout human history; even our closest relatives, the chimpanzees, go hunting. By their excited hoots and thrashing displays, they show how thrilled they are when they are successful. Hunting is at the heart of life for many animals. I wanted and needed to understand it.

Could I be a falconer? I weigh my options. On one hand, I have my hens, my home, my marriage, my job. On the other hand, my desire.

❧

On a freakishly warm, cloudy morning in October, Nancy tells me about yarak.

It's a word that is difficult to define, Nancy says. Its origins are obscure. The *Oxford English Dictionary* says it might come from the Persian *yaraki*, meaning power or strength, or from the Turkic, for the proper heat for tempering metal. Falconers speak of a bird being "in yarak"—in proper condition for hunting. But yarak is not about physical health or strength. Yarak is instead something central to the psyche of a bird of prey.

The bird who taught her about yarak was her first Harris's hawk, Indy, a gift from a good friend who breeds hawks and re-habilitates eagles. A big, strong bird, Indy was a powerful hunter from the start. This was back when Nancy had just finished her falconer's apprenticeship, before she and Jim had moved to rural Deering, where they live now. They were then living near a large housing development, where wild game was scarce. She offered Indy captive-bred live quail and partridges; in the fall, Indy also hunted migrating birds.

The winter started out mild. But then came a storm. The migrants left. And then Nancy ran out of captive quail and par-tridges.

But Nancy wasn't worried. She flew Indy frequently from tree to glove. He was eating plenty and getting exercise and com-pany; he just wasn't hunting.

Then one day she was with him in his mews, changing his

jesses. He reached out with his foot and circled her wrist with his toes—and held on. "He didn't hurt me," she said. "But he was dominating me. This is called 'braceleting.' But I didn't know it at the time." She finished with the jesses.

Two days later she made a mistake. She was changing the jesses again and unwisely changed the one farthest from her first, leaving the leg closer to her free.

Indy exploded. "It was as sudden as a windstorm," Nancy said. Without warning, the bird screamed at her and flew at her face with his talons.

He hit so hard she felt she had been punched in the mouth. He sliced her so deeply that fifteen years later, you can still see the scar: a thin white line running from her left nostril to her upper lip. "By the time I finished changing the jesses," she said, "the blood was flowing so hard it was covering my hands."

This wasn't crabbiness, like Banshee. This was an attack. "It was like he'd never seen me before," she said. "I think what he did surprised even him."

The attack made no sense to her. Harris's hawks are supposed to be cooperative, "but this was not cooperative," she pointed out. What was going on?

Her husband, Jim, with whom she had apprenticed, said it might not happen again. She asked the president of the North

American Falconers Association. "I don't know," he confessed, "but Indy might be as upset as you are." Then she called Frank Beebe, a falconer of international repute, author of many books on hawking. "I have a feeling you need to think back," he told her. It sounded to him like a buildup of yarak—the often-explosive buildup of hunting desire.

Nancy went over in her mind the events of the previous six days with Indy. It had all started, she realized, when she'd introduced him to game—and then cut him off.

"I had not realized it was like flipping a switch," she told me. "When you cut it off, it's like the power goes off and the oil burner floods. You have to turn the switch off before lighting it again, or BOOM! I hadn't turned the switch off. The hunting drive wasn't going off."

Yarak names this drive, this desire. It is the bird of prey's greatest earthly pleasure and its deepest frustration, twined tight. Frightening and beautiful, yarak is rapture and longing, love and death married in one timeless moment. If the eye is the mind of the raptor, then yarak is its wild soul. Yarak is wildness incarnate—dangerous and delicious and pure. Yarak names why I long for communion with birds of prey. This is as wild as wildness gets. There is but one way a human can touch a hawk's wildness: to join it, as its partner, on the hunt.

One day Nancy phones me at home. Usually we set up the next lesson at the previous one, or we communicate by email. The phone call feels momentous.

"I want you to see the end result of what we do," Nancy tells me. "Up to now, you've been taking baby steps. It's great for you, but very unsatisfying for the birds.

"The whole thing about falconry," she says, "is getting to hunt. Not to hunt with a falconry bird is to deny what they are. Hunting is the strongest instinct they have. It overrides migration. It overrides procreation. Hunting game is what cements the partnership. And you will be part of that successful hunting file."

She is inviting me to witness a hunt. Tomorrow.

"I want you to see in the field how these birds respond to everything I'm teaching you at the house," she says. "It's magic."

My heart pounds. I am hungry for this.

The day's hunt will involve, Nancy said, something called a quail launcher.

Immediately a shiver of horror goes through me.

Nancy asks gently, "Would that upset you?"

"No." I desperately want to go.

The next morning Nancy picks me up in her truck. In the

back, from their travel crates, the hawks Smoke, Sidekick, and Scout are shrieking like sirens, and from within a portable kennel, a female German wirehaired pointer, Stormy, whimpers with excitement. We are heading to a hunting club, which stocks its extensive fields with captive-bred pheasant and quail. Nancy outlines the plan for the day.

"What I'm trying to do," she says, "is first take Smoke, who already works with dogs. With Stormy and Smoke, we'll do one sweep of the field to see what the dog turns up. And then we'll do the quail launcher."

Two quail to be "launched" will be awaiting us in a box, courtesy of the hunting club's owner. The launcher is not, to my relief, some sort of cannon or catapult that throws quail. It's more like a folded-up trampoline. The act of being launched does not hurt the quail, although it is surely frightening. Almost immediately the quail begins to fly on its own.

So why the quail launcher? For the falconer, the contraption is a teaching tool to help a young hawk learn that the dog is part of the hunting team—which left to its own devices, the hawk would not discover. Harris's hawks naturally hate dogs. In the wild, they hate coyotes so much they scream at them. This is exactly what Smoke will do when she catches sight of Stormy—in the process alerting any nearby prey that a hawk is near and ruining the chance of a successful hunt. The hawk will never get the chance to see the dog put up game.

Using the quail launcher can help the falconer control the situation. Done correctly, the quail can be launched while the bird is looking in the right direction to see the dog on point. "Once the hawk realizes the dog puts up game, he ceases resenting the dog," Nancy said. "He'll start to take cues from the dog. These birds, their instinct to hunt is so strong, they'll never forget what constitutes hunting success—and now the dog will be included."

We are driving to the field where we'll ask Stormy to look for pheasant with Smoke. I'm scanning the sky and the treetops, as I always do—looking for hawks.

"Look who's here," I say to Nancy. There's a male redtail in a tall, leafless maple.

Nancy knows him. "I've had him hunt my birds," she says. She gets out and lets the dog out, too. "Go away!" she shouts at the wild redtail. As if obeying, he flies toward a pond, floating on the strong wind, circling—and then dives down, dropping like a stone. "He went down with a purpose," Nancy says. He saw a pheasant. There's game out here.

But finding it won't be easy. It's windy. The wind interrupts the scent the dog needs to follow. "This is a hard day," Nancy says.

The plan is to walk back to the cover at the end of the road, where pheasant might be hiding. Smoke steps out of her carrier. From Nancy's glove, she flies to a tree. She can better watch the dog from there.

"Stormy—hunt 'em up! Find bird! Find!" Nancy says to her brown-and-white dog. Stormy is glad to oblige. Nancy and Stormy have been hunting together for nearly nine years, ever since Stormy was "the puppy from hell."

"If it was something that would be featured on *Antiques Roadshow*, she'd chew it," said Nancy. "But that's what makes a great hunting dog—a smart, active, curious puppy."

Stormy's sniffing everywhere, zigzagging back and forth, searching for the edges of a cone of scent that will lead to the game. The large brass bell at her collar—both dog and bird wear bells to help us keep track of them—chimes with excitement.

"Stormy's getting birdy," Nancy says under her breath. And now the dog beelines. She's tracking now. And from her treetop perch, Smoke is watching, so intensely I can feel her gaze burning, hot as a laser. Smoke shakes her tail. It's a sign she is getting ready to fly—either to a new perch from which to better observe Stormy or after game she has spotted herself.

We're walking through fields rough with the dry brown seed heads of goldenrod and ragweed, the thorny stems of wild raspberry. The white fluffy parachutes of milkweed are strewn everywhere like feathers. Beneath the thick cover, there are holes to twist your ankles, and at face level, branches to poke your eye. But most of all, if you are carrying a bird, you must guard the safety of its eyes—not yours.

Filled with tension, we're all watching, trying to see everything at once. We're watching the dog's posture. We're watching Smoke's attention in the tree. We're watching for pheasant. And we're watching out for the redtail, should he reappear. We are not the only hunters here. Smoke could be killed, just like any wild hawk, just like any wild prey. This is a threat that all wild birds—hawk or dove, crow or pigeon—must live with at every moment.

Nancy wants to call Smoke to her glove, where she'll be safe. She takes a chicken leg from her pocket and holds it up to call her. But Smoke isn't hungry for food; she's hungry for chase. She is ruled by yarak. She keeps her eye on the dog.

Now the tinkle of Stormy's bell stops. Just above the ragweed, we can see her spotted back go rigid. "Look what we've got!" cries Nancy. "A point!" But a second later: "Oh, damn!" Sure enough, the redtail's back. He soars above us; soon it's evident that this hawk, too, has seen what Stormy has pointed out.

The pheasant bursts from cover. "Get it! Get it!" Nancy cries urgently. Smoke flies after it. She misses and lands in a tree.

In her short lifetime, Smoke has not yet killed a pheasant,

An eager hunting partner flies to master falconer Nancy Cowan's glove.

but she has chased three. Her first time was just three weeks ago, right here. Smoke was on Nancy's glove. Right in front of her, Stormy got a point, a hen pheasant flew up, Smoke chased it. And from that moment on, Smoke understood that the dog was an integral part of the hunt.

Now the pheasant flies toward a distant pond, and the redtail flies after it. Smoke wants to follow, too. But the redtail is a real danger to her now. Smoke takes off.

"Smoker!" Nancy yells at the hawk, furious, frightened.

"Don't go there! I know the pheasant went there . . ." She blows the whistle. Smoke circles, lands in a tree, and looks down at us, disgusted. "She's mad she missed that pheasant," says Nancy. We'll try for another. "Stormy!" she commands. "Hunt 'em up!"

But the dog, too, is frustrated. She can't find another scent. She heads toward the tree where Smoke is perched.

"NO!!" Nancy yells. She knows what is happening: Sometimes if a dog can't find a pheasant or grouse, it'll go for the nearest bird—the hawk. And that makes the hawk furious: "Don't you point me, you moron!" Once Stormy pointed Jazz, and the hawk, enraged at the dog's stupidity and overcome by yarak, shrieked in fury and frustration and flew at Stormy with her talons. Nancy had to extract Jazz's feet from the dog's snout.

It seems inevitable: It will be a bloody, dangerous day. Yet I am not disgusted or distraught. I am living through my eyes now.

We've been searching for a good twenty minutes when, at 10:00 a.m., Nancy announces, "We've got a point!" Again Smoke is watching the dog intently. "She knows Stormy is birdy. Sometimes the bird sees 'em and wants the dog to flush 'em. Eventually they figure out a way to communicate that," Nancy says.

We're tromping over rough territory. Later I find that my pants are torn, my legs scratched and bloody. We're both sweating with exertion on this cool late autumn day to keep up with dog and bird. "You see why it's called 'hunting' and not 'getting,'" says Nancy. "You are seeing the essence of the hunt. Even if she doesn't get the game, the slip on game"—the chance to chase it— "is valuable."

And that is all Smoke will get today—a chase, but no catch. Nancy lets her attack the baited lure to satisfy her yarak and puts her back in her carrier. "Raptors are never happy," says Nancy. "They are always wanting, wanting to hunt. And if they are full, they are just flatlining." They are content, perhaps—but really, it is like their souls are in storage, awaiting the next chase.

At 10:50, we go to pick up the two quail that the owner of the hunting preserve has set out for us. He has left them in a little pet carrier in the garage. I look inside and my heart melts. There is nothing more innocent and appealing than a quail, with its rounded profile and soft brown plumage and black button eyes. One of the favorite books of my childhood was *That Quail, Robert*, about a quail who lived in a house with her people. Robert the Quail was so well loved that when the wife's diamond popped out of her setting and Robert swallowed it, the couple picked through quail droppings for weeks rather than sacrifice the bird. (Robert, in fact, kept the diamond, as grit to grind food in her crop.) I love the dear question marks on the heads of wild quail in Arizona, curled crests of black feathers. Watching quail chicks parade after their mother, in single file through the grass, I had been overcome with anxiety for their safety.

And now, I've come to watch these quail be killed. In their pet carrier, the two birds are still as stones, their stillness a fervent prayer that we somehow won't see them.

At least the quail launcher looks less terrible than I'd expected. It's an oblong box made of folded cloth, activated to spring open by remote control. Nancy positions the launcher carefully, then enfolds a quail inside like an astronaut in a space capsule. Stormy is ready, eager. Scout is still in his carrier in the car. But something goes wrong. The electronic launcher goes off prematurely and the quail launches to freedom. The little bird flies away and disappears into the brush before hawk or dog sees it.

I am flooded with relief. And then I feel a pang of guilt. Whose side am I on?

We have one quail left. My heart is pulled in two directions. Nancy places the little bird in the launcher. Stormy rushes over to smell it. The dog could trigger the launcher and release the quail before Scout is even on the glove. "Bad dog!" Nancy yells. Stormy comes to her side, chastened, and I hold the dog by the collar until Nancy has everything in place.

"WRAAACCCK!" Scout cries. Now he's on Nancy's glove. He's in yarak. He wants to hunt. He hates the dog.

"WRRRRAAAACCK!" he screams at Stormy: "YOU STUPID DOG! I HATE DOGS! I HATE YOU, HATE YOU, HATE YOU!" He shakes his tail, wanting to fly.

"We'll let Stormy go in and point," says Nancy, "and then Scout will notice something more important than hating the dog is going on."

But after Nancy had yelled at her, Stormy understandably doesn't want to go anywhere near the quail launcher now. The dog is everywhere but where Nancy wants her, obviously avoiding the device. "WRACKK! WRAAAAACK!!" calls Scout. He's extremely agitated. I can feel the buildup of frustrated desire. I worry he might bite or foot Nancy.

Now the dog is "getting birdy"—but not at the right bird. Stormy has found a pheasant—and wants to follow its scent instead of pointing at the imprisoned quail. "Storm-a-thon!" Nancy calls, urging her toward the launcher. "There! Go there!" Finally, the dog points. Scout is watching, and his attention is palpable, hot as a laser beam. Nancy presses the red button on the remote control. The quail shoots fifteen feet into the air and starts flying.

In a split second, Scout rockets off the glove, flips upside down, and attacks the quail from beneath, grabbing its belly with his talons. "Oh!" I cry. At that moment, there is no room in my soul for the quail's pain and fear. I am flooded with the hawk's elation. I feel it like a drug in my bloodstream, the ancient thrill of hunting success. I have never felt this before, yet it feels as familiar as my own skin. Through this bird, I have touched something very old and very wild, something I thought I could never feel: yarak. I have no desire to hurt that quail—but I realize that I want, more than anything, for this hawk to catch it. In the process, I have also learned something important about the nature of love. Now I understand why my father, an animal lover, bought my mother all those fur coats: his love for her blinded him to everything but her fire-bright happiness.

"You did it, Scout! Good bird!" Now he's got it; he knows how to hunt. Nancy and I are proud. But as Scout flies off with his prize, our elation turns to anxiety. We could lose him. Indy once caught a chukar partridge here, disappeared with the prey, and wasn't seen for three days. We race after the bird.

Scout has flown toward the forest, on the other side of a hedgerow. If he has landed, we can't see where. And now overhead, against approaching storm clouds, another soaring shape draws Nancy's attention.

"Oh, no," says Nancy. "Is that a gos?"

"It doesn't look like a redtail," I say.

Carrying the quail makes Scout's own body an easy and attractive target for a more experienced hawk. Even though a gos is smaller than a Harris's, "whoever is lower than you is easy prey," Nancy reminds me. It could easily kill Scout.

"Hold the dog!" Nancy says. If Scout has landed on the ground, the dog might scare him off. I hold the dog by the collar while Nancy races ahead. I'm grateful to remain behind. We know that Scout has caught the quail, but we don't know that he has killed it. In her pocket Nancy carries a sheathed knife for such an occasion: if the hawk is eating the prey but it is not dead, to spare its suffering, she will cut the head off the living quail.

⌒

I still visit Nancy's birds and fly them from time to time. With my traveling schedule, I can't possibly take on a falconry apprenticeship at this point in my life. And then there's the not-inconsequential matter of my marriage and our hens. But Nancy and her hawks have profoundly changed me. Now, even as I eat my broccoli and vegetable lasagna, even as I pray and work for compassion toward all sentient beings and write my checks to humane causes, I understand the falconer's mantra. People do not have to hunt, but hawks do; and these words name the unspoken rules by which hawks everywhere live their innocent, incandescent, wild lives:

"Hunt hard. Kill swiftly. Waste nothing. Offer no apologies."

SELECTED BIBLIOGRAPHY

Hawks

Bodio, Stephen. *Eagle Dreams*: Searching for Legends in Wild
Mongolia. Guilford, CT: Lyons Press, 2003.

——*A Rage for Falcons*: An Alliance Between Man and Bird.
Boulder, Colorado: Pruett Publishing, 1992.

Clark, William S. and Brian K. Wheeler. *A Field Guide to Hawks
of North America*. Boston: Houghton Mifflin, 2001.

Ferguson-Lees, James and David A. Christie. *Raptors of the
World*. Boston: Houghton Mifflin, 2001.

Parry-Jones, Jemima. *Falconry*. Devon, UK: David & Charles,
2003.

General Ornithology

Barber, Theodore Xenophon. *The Human Nature of Birds*. New
York: St. Martin's Press, 1993.

Gill, Frank B. *Ornithology: Second Edition*. New York: W. H. Freeman & Co., 1995.

Kilham, Lawrence. *On Watching Birds*. Chelsea, Vermont: Chelsea Green Publishing, 1988.

Lorenz, Konrad. *On Aggression*. San Diego: Harcourt, Brace & World, 1966.

Skutch, Alexander F. *The Minds of Birds*. College Station, Texas: Texas A&M University Press, 1996.

Stokes, Donald and Lillian. *Stokes Field Guide to the Birds: Eastern Region*. Boston: Little, Brown, 1996.

——*Stokes Field Guide to the Birds: Western Region*. Boston: Little, Brown, 1996.

Tate, Peter. *Flights of Fancy: Birds in Myth, Legend, and Superstition*. New York: Delacorte Press, 2008.

Also highly recommended*

Cowan, Nancy. *Peregrine Spring: A Master Falconer's Extraordinary Life with Birds of Prey*. Guilford, CT: Lyons Press, 2016.

Golding, Philip. *Falconry and Hawking: The Essential Handbook*. Oxford, UK: World Ideas Ltd., 2014.

Macdonald, Helen. *H is for Hawk*. London: Jonathan Cape, 2015.

* *These excellent books were published after the experiences in this book were lived and recounted.*

Websites

Global Raptor Information Network lists organizations
working to conserve birds of prey: https://www.global
raptors.org/grin/siteLinks.asp?lid=1000

To observe the hawk migration in the United States, Canada,
and Mexico, consult this map maintained by the Hawk
Migration Association of North America: https://hawkcount
.org/sitesel.php

Track Monadnock and the other tagged hawks in the Broad-
winged Hawk Project: https://www.hawkmountain.org
/conservation-science/active-research/raptor-conservation
-studies/broad-winged-hawks

ABOUT THE PHOTOGRAPHER

Originally trained as an oil painter, Tianne Strombeck now creates photographic portraits of nature to promote conservation and wildlife education. Her painting background has given her an instinctive understanding of color and composition. She works to understand her subjects, to capture their essence, and to show how they interact with their environment and one another. Her work has appeared in numerous books, including Sy Montgomery's *The Soul of an Octopus*, *Condor Comeback*, and *The Hummingbirds' Gift*. To see additional examples of her work, from hummingbirds to jaguars, visit her galleries at https://www.tianimal.com.